Balance

by Sheila Rivera

first step nonfiction

Lerner Publications · Minneapolis

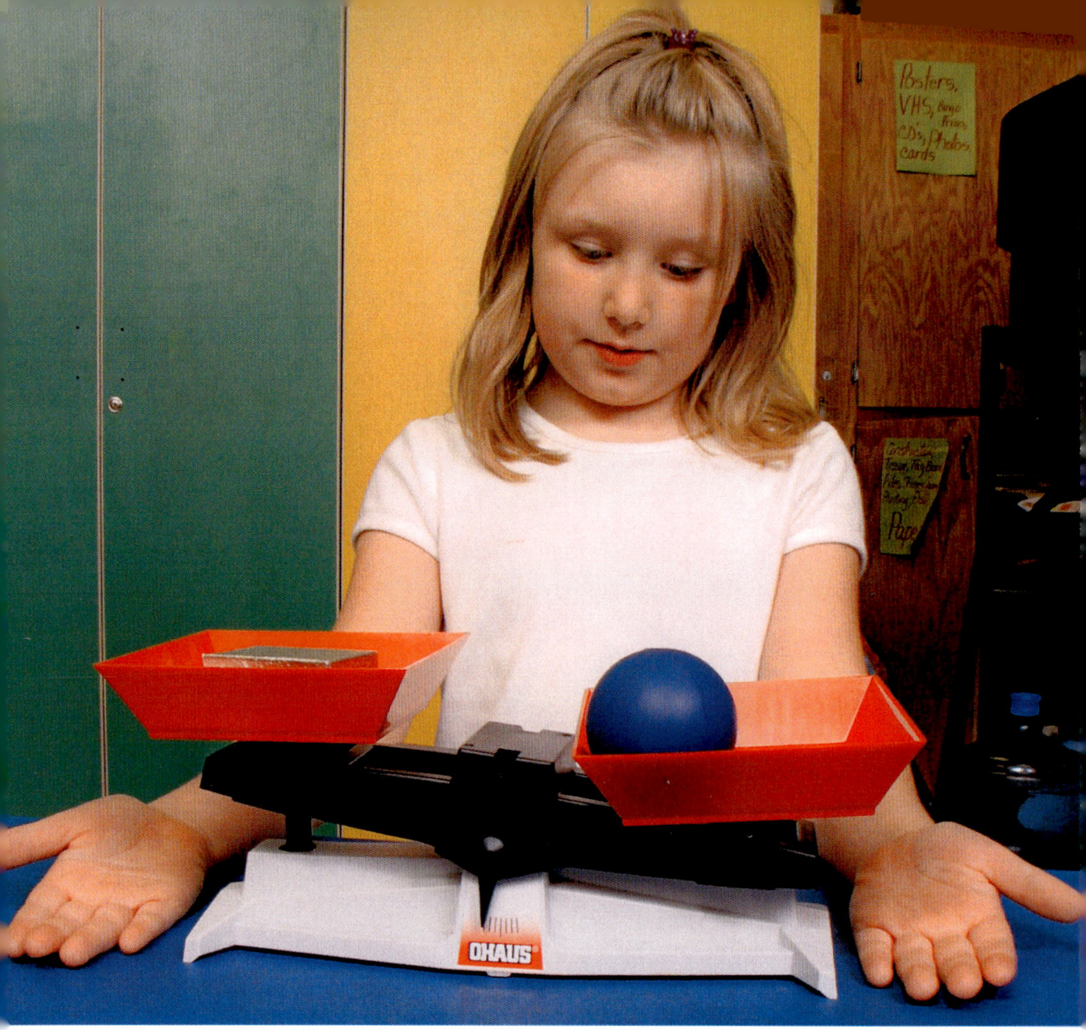

I use a balance to
weigh things.

A shell weighs more than a feather.

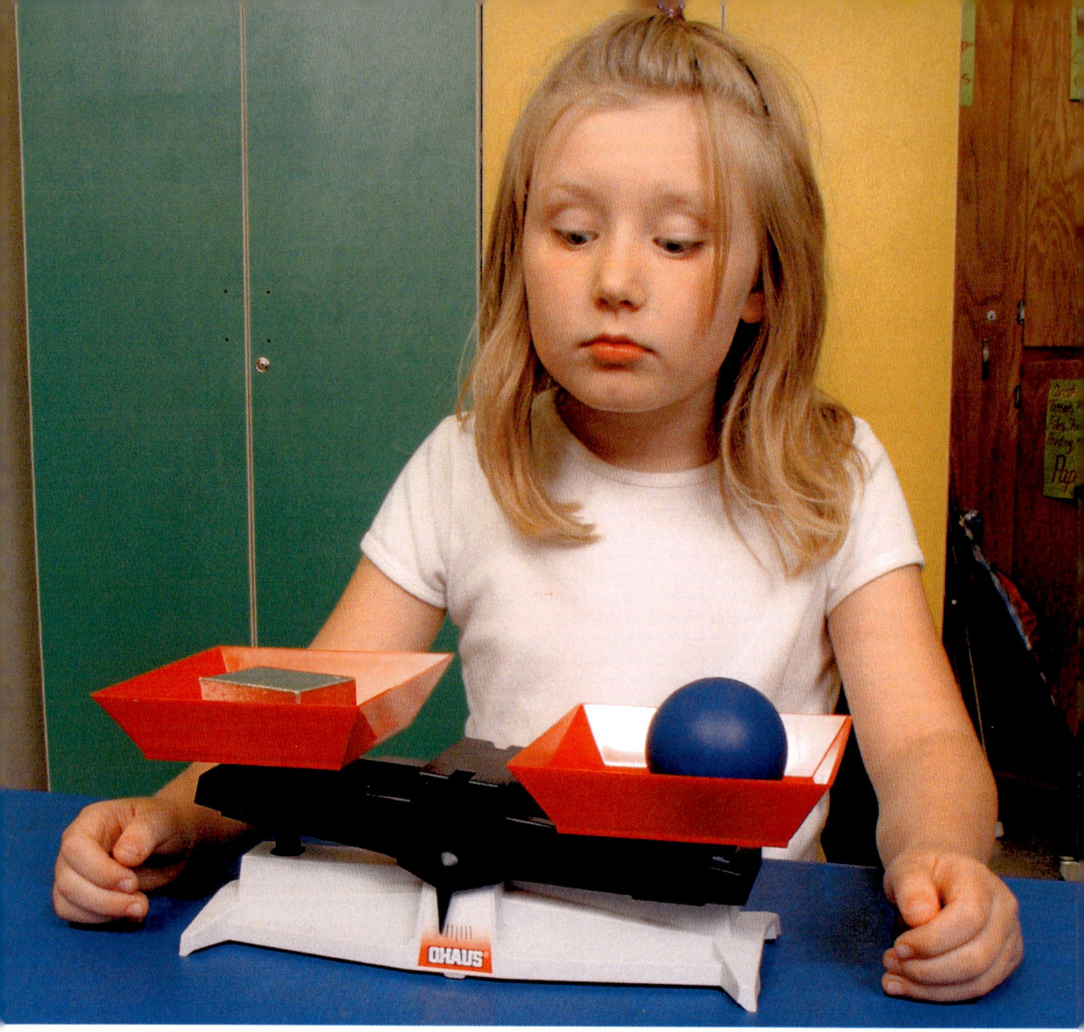

A box weighs less than a ball.

A leaf and a feather
weigh the same.

Which one weighs more?

Which one weighs less?

Can you use a balance?